The Ancient Tradition of Beekeeping

Bees in Your Backyard

Dueep Jyot Singh

Farming Series

Mendon Cottage Books

JD-Biz Publishing

Download Free Books!

http://MendonCottageBooks.com

Our books are available at

1. Amazon.com
2. Barnes and Noble
3. Itunes
4. Kobo
5. Smashwords
6. Google Play Books

Download Free Books!

http://MendonCottageBooks.com

Table of Contents

Introduction

When I tell you offhand to give me your immediate response the moment you hear the word Honeybee, your response is going to be "honey," about 99.9% of the time. And then after that, you are going to say "beeswax." But the product of a beehive is not restricted to just these two items. It also has royal jelly, pollen, and *propolis.*

All of these products have traditionally been in use by mankind for the propagation of good health, beauty, and products in use around the house and as a health food for millenniums, going back to prehistoric times.

So this book is going to tell you all about how you are going to be introduced to bees, and if you are enterprising enough to keep them in your backyard, you are soon going to be called an apiarist, keeping a number of your hives in your apiary.

I was visiting a prosperous family farm passed on through generations, belonging to some Greek friends and was admiring their poultry, their ducks, their livestock, and their horses, when I just happened to say, "Where are your bees?" And I got a look of surprise.

Naturally I returned it back because I was under the impression that beekeeping methods have come down the ages in older civilizations in Europe, especially when everybody knows that in ancient Greece agriculture it was a very highly respected farming tradition.

In fact, even today archaeologists have found bee extractors, bee smokers, and other beekeeping paraphernalia going back to prehistoric times.

That is because so many people have begun to forget the traditional art of beekeeping, because of lack of knowledge. Also, I do not think that there are many training centers in highly developed countries where sericulture is taught to the local populations in order to add significantly to the family kitty in terms of financial prosperity.

That is because bees have been given bad PR. because of people who are ignorant, or through rather stupid reasons, because a Kenyan friend told me that her village stopped beekeeping about 300 years ago, because about 300 years ago, some people from Europe came there, and decided that this was a heathenish practice, especially the giving of honey as a bride price and the drinking of fermented honey in celebrations was sinful and unacceptable in the eyes of God!

Believe it or not, this is true, and has happened throughout Africa, since the 17th century because in many parts of Europe, ignorance about beekeeping meant that there was some fool somewhere who would prevent this tradition from being practiced by an indigenous people because he said so.

This is how beekeeping practices in Kenya, Ethiopia, and even Egypt, where beekeeping was traditionally practiced as early as 3000 BC, if you go by the Egyptian wall murals, was eradicated by unaware and ignorant people who wanted to impose their own notions of civilized practice on an ancient people.

Traditional beehive in Ethiopia

However, in many parts of the East, where such people are not given due acknowledgment and their interference in the local tradition and culture was not encouraged, even today sericulture is a way of life. For example, in many parts of Asia, villages are known by the quality of bees and the honey produced by them.

They keep those bees in their backyards, and it is the job of the women of the family to take care of the bees.

My favorite uncle and aunt had 12 hives, which they kept on the terrace of their house, and not in the garden because their pet labrador, Oscar, had this curious habit of inspecting the garden from corner to corner every single day, twice a day without fail.

But he was not allowed onto the terrace to his great indignation. So one fine day, when he was showing his indignation because his dearly beloved Master disappeared upstairs and locked the door for about 3 to 4 hours every day, uncle decided to give him a talking to.

Incidentally people who keep dogs hold long conversations with them and the conversation went something like this, to our great amusement.

Uncle – all right then Oscar, if you are really so curious, you can go up and inspect whatever is there.

Oscar – wagging his tail – oh thank you, thank you, thank you so much. I really was so curious about that area upstairs, and I will be back soon, do not you worry.

Uncle – but be careful, Oscar, curiosity killed the cat. Do not come yelping to me when you find yourself stung.

Oscar – Who, me, master, who could harm me, master? I am the most well behaved dog in the neighborhood, and the whole World is my friend.

Less than a minute later, we heard a series of yelps, and there was Oscar scooting down the stairs to go creeping under the bed. He had gone and stuck his nose into a beehive.

He had to be coaxed out so that we could remove the bee stings – three – from his little nose and apply some honey on that poor little mistreated, inflamed organ. And uncle kept on, "I told you so, did not I, my baby, but you have to be shown things!" And Oscar muttering – "oh shut up…"

Incidentally, after this incident, Oscar became quite good friends with the bees. Everything was forgotten and forgiven, because the beekeeper told the bees that Oscar was a valuable member of the family and introduced him to their beehives!

I found this whole procedure fascinating, and uncle told me that this is traditional bee lore. When you are keeping bees, you have to tell them everything which is going on in your family, including births, happy occasions, and even deaths. Otherwise they are not going to consider themselves part of the family, and they are not going to return to the hive again next season.

I used to go on their terrace, very often, and I was never stung, because somehow the bees knew that I was not afraid of them. Nor would I harm them ever. One day, I came down from the terrace, and my cousin shrieked - she was terrified of them – that there were some bees on my jacket and I had better hit them or just fling them away from me.

Instead, to her great surprise, I said they are not going to do anything to me, went into the garden, leaned against the nearest flowering tree, and the bees calmly walked from my jacket to the plant, humming away gladly. My cousin looked at me with acute dismay and scowled, – "do not tell me that they are crazy after you too, sis, they are crazy after mom and dad, Oscar, grandpa and everybody else, except me, so what is the explanation?"

The explanation is, as I said bad PR.

Down the centuries, bees have been equated with their really bad tempered cousins, the wasps and the hornets. These insects do not mind stinging you often, and without any provocation whenever they want.

On the other hand, a bee is not going to sting you unless of course it is protecting the hive or the Queen bee. Also, once a bee stings you, it dies. That is because the stinger is left in your body, and when after the bee has stung you, the stinger is pulled away from its body, literally. So the poor little bee sacrifices its life in order to save its family, its Queen, hive, and tribe.

No wonder, in Lithuania, the Bee is considered to be a symbol of peace. And if you want to praise a friend or do him honor you are going to call him a friend and a brother through the kinship of bees *–biciulyste*. The Lithuanians also had a bee goddess.

History of Beekeeping

Once upon a time, mankind collected honey from honeycombs, in the woods, rock croppings, and other places where he could get honey easily, with the help of just hitting the honeycombs with some missiles, standing on the ground and with some protective covering, covering him from the attack of the bees.

More than 10,000 years ago, he decided that it was easier for him to keep the bees at home, in clay jars and vessels. Archaeologists have found the vestiges of honey in the tombs of the pharaohs, – especially that of the most famous archaeological discovery of them all, Tutenkhamun. The ancient beekeepers knew how to smoke out the hives when the harvesting needed to be done and this was done with smoke in order to stun the bees and make them so dizzy and drowsy that they did not bother to sting the collector of the honey because they were so busy getting their senses back to normal.

At that time, alas, the whole honeycomb had to be destroyed in order to get access to the beeswax and to the honey. However, as time went by, people began to know more about beekeeping ways and methods where these honeycombs could be raised in wooden boxes, natural hollow logs, and also in straw baskets, as is the practice even today in the Middle East.

Archaeologists have found remains of 30 beehives dating back to the bronze as well as the iron age, in the Valley of Jordan. These are made of unbaked clay mixed with straw. This proved that about 3,000 years ago, and even farther back than that, there was a thriving honey industry in that area, with more than 1 million bees in about 100 hives set out in rows. This annual yield would be more than 500 kg of honey and about 80 kg of beeswax.

Even Aristotle discussed the life and the times of bees and how they were kept, along with other Roman writers and historians like Columella and Virgil.

In ancient China beekeeping was an art, which was described minutely In Business Success Books by writers. The quality of the wood used in the protection of the beehives, as well as the experience of the beekeeper were all described millenniums ago, talking about how these bees were kept during the autumn and the spring season.

On the Indian subcontinent, beekeeping was an ancient art, which was done with five varieties of bees. Even today, there are writings about forests of honey, where the tribals had been raising wild bees, down the centuries, and they know how to collect the honey at a given period of time. When I was young, I often found these huge honeycombs on branches in the jungle, and because I left them alone, they left me alone.

Also, we knew that there were a variety of bees that were very very aggressive and dangerous, and they were the rock bees. They were capable of stinging local hairy bears, wolves, and other animals to death and we were just simple ordinary human hair-less kids, so we kept far away from them.

The ancient Mayan civilization was lucky, because there was a variety of native bees in their area which did not sting. They were kept by the Meliponini tribe, and even today the culture of these bees in Brazil is still a major part of the economic cottage industry and has also spread to Australia where the stingless bee *Tetragonula carbonaria* is raised in what is called meliponiculture instead of sericulture or apiculture.

The term "beekeeping" is not restricted to just the care and housing of bees. It also talks about all the techniques, which you are going to be using in keeping these bees, harvesting the products which you obtain from their beehives, and the further processing of those products into items which are useful to mankind.

Incidentally, if there were no bees in the World, you would not have a number of flowers species growing or existing today, because there are plenty of flowers and plant species which can propagate further only upon pollination by insects like bees. So consider these bees to be your best friends helping in the next stage of the apple harvest.

Getting the Beehives

Believe it or not, if you ask a beekeeper to move his beehives from one area to another, he is immediately going to shout, Absolutely Not, and anathema Maranatha. If you think that they are being very possessive about giving their beehives to somebody else, or even sending them to another country or region, he is going to tell you that a number of bee diseases and pests have appeared in beekeeping yards, because of this transfer of beehives from other places, bringing their disease bearing pests and parasites along with them.

That is why in a number of countries, the ancient beekeeping industry has gone bankrupt.

This spreading of the disease began in the late 70s, and today, nobody is going to start up a beehive, bought or brought from somewhere else, especially from abroad. They are going to use the traditional ways in which a colony nesting in the wild can be enticed into an already made beehive in your farm or on your garden.

The colony already has a number of honeycombs. You are going to transfer one of these honeycombs from this colony into your hive. This honeycomb is now going to be tied into the frames of the hive, so that the bees know that the support is steady. You can also tie the comb onto the top bar of the hive.

If you want to entice the bees, you would want to place the hive in an area where there is a wild colony nesting or you know that bees are swarming. Rub the inside of the hive with honey and beeswax. Here is one indigenous solution, which was used for millenniums.

Bees love the scent of Oscimum sp - sacred basil – so get some twigs of sacred basil, and dip them in beeswax. Then light these "smudge sticks" and whirl the smoke all around the hive. The bees are going to be enticed by this scent and say *hey, here is one ready-made hive all waiting for us, let us start building a colony right here, whistle while you work.*[1]

However, this is going to be possible only where plenty of bee colonies are in existence, especially when you are trying to entice those bees into your

---

[1] Incidentally, apart from humming, I have heard a number of bee species emitting a variety of different sounds, and sometimes I thought I even heard them chirping musically. Maybe they were just happy or they were just gossiping away.

backyard. You may also want the assistance and advice of local experienced beekeepers to give you a queen bee and a part of a good honeycomb, to start up your colony.

Placing the Hives

Having traveled extensively, I have seen traditional ways and methods in which bees have been raised, and believe it or not, people have not bothered about special areas and land, in order to set up a beehive or a row of beehives.

In villages in Nepal and Burma, I have seen them placed on the rooftops and even on wasteland, which could not be used for farming. They are also placed on a number of large trees, and funnily enough, the bees know that these trees are held sacred, and they do not mind swarming there, even though occasionally human beings do turn up to plunder their honey combs.

The area should just have two factors – it should be well protected against direct sunlight. Bees do not want to get roasted, especially in hot weather conditions like in Ethiopia, Kenya, and other African countries. Also, if there is a high Gale blowing or the area is windy, the bees are not going to swarm there, because they do not intend fighting a Force Two wind every day when they return to their hives.

Any sort of container can do, with an area where your honeybees can make the hive, and nest. In prehistoric times, open mouthed, wide clay containers were used for these colonies. Today, apart from traditional hive structures, you can also have hives with top bars and hives with movable frames.

A Traditional Natural Hive

This is where you are going to be using some of your creativity, and go back to the wisdom and the experience of the ages, in order to use natural and local indigenous materials in order to provide your natural hives. Apart from clay pots, woven baskets, cane baskets, and logs which have been hollowed out, you can use anything which is hollow and which is sturdy enough to support a bee colony. Remember that you have to make that colony accessible to you, because at the time of harvesting, you should get easy access to that hive and the products of the honeycomb.

It is going to depend on the skill of the beekeeper whether the bees are harmed during the process of harvesting the beeswax and the honey. In ancient times when there were plenty of honeycombs available in the wild, people did not bother much about the survival of the bees because during the next swarming season, they would get access to another swarm easily. But nowadays, with a number of bees species coming into the endangered list because of diseases spread globally, you have to be very careful about the possible killing of the bees during honey harvesting.

Remember that if you destroy the hive, the bees are never going to come back to the hive again, next swarming season. That is why in many parts of the World, even today, after the honey is harvested, skillfully by an experienced beekeeper, part of the honeycomb is still left in the hive, reassuring the bees that they can build their honeycomb in that particular hive again. Remember that if you have a number of hives, do not expect all of them to be occupied by bees all the time. They may decide to go and occupy your neighbor's empty hive, because you destroyed the colony a bit too ruthlessly during the harvesting, and bees, like elephants, never forget.

Top Bar Hive

A top bar hive was supposedly a relatively new modern innovation until it was found that it had been in use in Africa for more than 3,000 years. The bees are going to be encouraged to make the honeycombs onto the under-side portions of a number of bars. That means the beekeeper has easy accessibility to all these honeycombs.

All he has to do is just lift up the bars, and he is going to get all the honeycombs moving away in his hand, depending on the individual particular colony, that he wants to inspect right now. You can use any available local material in the construction of these top bar hive boxes.

Here is one easy way in which you can find out the distance between the bars, when you are constructing this sort of hive frame box. This is going to depend on the species of the bees that you are keeping in your backyard. African Apis mellifera – width between the top bars, 34 mm. European Apis mellifera – width between the top bars – 35 mm, and Apis cerana – a common variety found in Asia – 32 mm.

If you can manage this, just measure the spaces between the honeycombs, in a wild colony nesting in your area. You are going to see how far the honeycombs are made in the natural state. Apart from this, you are going to see the volume of the artificial box/nest which you are constructing. It has to be the same as that cavity, which is normally used by wild bee colonies in the national state.

In ancient times, beekeeping in many parts of the World was supposed to be an activity to be followed only by men who could go into the woods safely to take care of their swarms. Their womenfolk, if they did any sort of beekeeping made sure that the bees were kept nearer to home.

Thanks to the introduction of top bar hives into the beekeeping industry, more and more people have found that this is a really inexpensive way, in which they can keep a large number of bees, especially for the honey as well as for pollination. These are very popular in Kenya, and in other countries in Africa where people keep honeybees only for social, cultural, and traditional reasons and not for selling them. Once the honey is extracted, the honeycomb is not going to be returned back to the original hive.

The third type is known as Movable-frame hives

Movable Frame Hives

Movable frame hives can also be considered to be a relatively modern design, being used in a number of developing countries in Asia, South America, and Central America. You are going to be using plastic frames are rectangular wooden frames so that these combs are supported properly. The advantages of a movable frame is that you can manipulate the colonies and inspect them by removing the frames and looking at the colony properly and minutely.

You can also move one frame from one stronger colony, to a weaker one, by just picking up a movable frame and taking it to the weaker colony so that it can be populated by the more densely populated bees species.

The best advantage of a movable frame is that you can empty out the frames of its product like honey and then you can return it to the hive. That means that the production of the honey is going to be increased more, annually, because the bees do not have to go through the process of building a hive again.

A frame hive is going to be made up of a number of boxes. Wood is of course the best material. Frame hives are excellent in sustainable farming, because you are going to be piling up the boxes, one on top of the other in a limited space. All the frames are going to be arranged in such a manner, that on first glance, you can think that you are filing a number of files in a cabinet. The bottom box is going to be used as a brooding nest. The Queen is going to lay her eggs there, and the larvae are going to develop properly in that nest.

A good movable frame hive is going to have a holed metal grid at its entrance. The worker insects can pass through that the grid but the Queen cannot do that. You can consider the Queen to be kept some sort of prisoner in her bee palace.

This metal grid is placed between two boxes. The brooding box, and the box placed above. That means the honey is going to be stored away in the boxes which do not have the Queen. The only downside to this particular frame hive is that you need a number of equipment pieces, which are specialized and other materials like a roof, floor, and a honey stand.

Many frame houses are normally constructed in Central America, North America, Europe, and Australasia, but I would suggest making them only if the cost of maintenance of these frames is relatively inexpensive in your country. After all, you have to maintain a precise design, with adequate spacing between the different frames which should be equal to that of a natural nest.

If you are a DIY type of person, you are going to find this URL very interesting, on how you can make a honeybee box –

http://www.wikihow.com/Make-a-Honey-Bee-Box

Fixed Honeycomb

Fixed honeycombs have been used down the ages in many countries where the main job of the beekeeper was to find the best way in which a beehive colony could be encouraged in his own backyard, with all the available, inexpensive material, which could include a hollow log or even clay pots.

You can find these fixed honeycombs in Asia, Africa, and South America.

Beekeeping Equipment

Traditionally, people made do with whatever was available to them, or use their own creativity and ingenuity in order to keep bees in a trouble-free manner. However, there is some basic equipment, of which you need the knowledge, to become a successful beekeeper.

Smoker

One of the traditional pieces of equipment which comes in very handy to an experienced beekeeper is the smoker. This is used to tranquilize the bees when the harvesting is done. Traditionally, the material used as natural fuel for smoking in the bees was dried cow dung which has been put in of your box, and to which the village blacksmith's bellows have been attached. Even today, I have seen this being used in villages to puff smoke gently into the hive. Apart from dried cow dung, other natural materials which can be used for smoking bees are cobs of corn, jute, and hessian, but never ever use tobacco smoke because it is very poisonous. The beekeeper is going to puff that smoke at the beehive's entrance. After that, the entrance is opened and bees are going to move from one section of the hive to another.

I have seen coconut fiber being used in many parts of Asia, especially in tropical regions where coconuts flourish and abound. Even today, traditional methods are to use any available non-odorous material which produces plenty of smoke and does not harm the bee.

Incidentally, the moment you use a smoker, you are going to tranquilize and calm the bee. For him it is going to be a natural state when the hive is on fire and that is why he is going to begin feeding on the honey. This is going to distend the stomach. These bees supposedly cannot bite under such circumstances – this is hearsay and has not been proven, – but it has been well known that the smoke is capable of masking the danger to the hive pheromones which are emitted by the guard bees, especially when some of the bees are accidentally squashed during the inspection or the honey removal process.

But you are never going to use smoke when you have a swarm. That is because they do not have a brood there, nor do they have honey. So when you are doing the inspection of a swarm, you do not need to smoke out the bees.

Protective Clothing

Naturally, you are going to need some protective clothing in order to protect yourself from possible bee stings, when the harvesting process is being done. My uncle used a traditional poncho design – once I asked him half seriously, why he did not use canvas or tarpaulin as a clothing protective material, and he said too heavy and bulky! – to protect his body, as well as a veil and a hat in the initial stages, but later on when he began to gain experience, he discarded this unwieldy protective clothing. He just began to use typical factory overalls because not only were they protective, but they were not bulky.

Also, he remembered to tie up the mouth of his sleeves and the legs of his trousers with elastic so that the bees could not crawl up his legs or his arms. The covering clothing is light in color. It is going to be smooth so that the bees do not confuse it with the naturally protective fur of their natural opponents like bears and other animals. That fur is dark, and the bees are naturally programmed to attack anything near their honey comb which has a dark rough fur and goes sniffing around the best way to gain access to that feast inside.

Also remember that the suits have to be washed regularly, because any sort of sting remaining in the cloth is going to keep pumping out pheromones. So the bees are going to attack immediately, the moment they smell these pheromones on your protective suit. You may also want to wash these clothes as well as your hands/gloves in vinegar solution to reduce any chances of pheromone induced stings.

Hive tool

There are a number of bee types and species, which are going to seal or close every joint or gap, between areas in the hive, with a product called propolis. This is also an important byproduct which you can obtain from the honeycomb.

To harvest this propolis, traditionally, people used knives, but they were flexible and that is why special pieces of metal which were not so flexible and could give proper amount of leverage to the beekeeper when he was removing the beeswax, separating the support and the frame ends, and also separating the different boxes were used very commonly. This is known as a hive tool.

Here is an excellent URL, telling you more about the hive tool.

https://www.youtube.com/watch?v=L3mEgjf2xpI

The Life of a Honeybee

Nectar being converted into honey

For thousands of years, mankind has acquainted himself with about 19,000 different species of honeybees, all over the world, down the ages. Most of them are wild and like a solitary existence. Some of them are giants and some of them are very small. Some of them are bumblebees and some of them do not sting. Some of them nest in the ground. A typical bee colony is going to have anywhere between 80,000 - 200,000 bees in it, depending on how strong the hive is and how old it is. In America and Europe, the species which is normally cultivated is Apis mellifera. In Asia Apis Cerana is more prevalent.

Nowadays, beekeeping researchers are experimenting with hybrids, especially to find out species which produce larger quantities of honey, are disease-resistant, and are of a milder temperament and behavior. However, it has been found out that these hybrids go back to nature in a number of generations, and become typically aggressive and naturally defensive.

One of my Egyptian friends told me a very interesting story about the friendship coming down the ages between a bird which is called a honey guide and the tribesmen collecting the honey. A honey guide would guide you to a honeycomb, you would smoke it out, and then collect the honeycomb.

And then to thank your bird friend, you would give it and its family a part of the honeycomb and thank it for being such a good friend. However, if you are selfish, and you did not give a part of the feast to the honey guide, the next time it chirped around you, saying, come on, follow me, my friend, it would lead you to the nest of the most poisonous snake it could find like let us say a mamba!

So that is why traditionally if you have a honey guide around, make sure that your good friends with it and share your feast!

In ancient times, of course, the harvesting was done by very crude methods because your priority was to obtain the honey and not bother about the destruction of the honeycomb. That honey was eaten immediately so nobody bothered much about the natural products in it like the larvae and other products of the honey comb.

But as time went by and bees began to be domesticated, everybody began to understand the value of the Queen bee who had to be preserved. And that is why in Europe, monks began experimenting on how to keep bees and harvest the honey without any sort of destruction to the hive, in their monasteries in the 15th and 16th century in Europe, especially in France, Germany, Austria, the Scandinavian countries, Spain, and Italy.

Every hive is going to contain just one queen. She is the queen mother, laying the eggs. All the workers are female, and the offspring of the Queen. So are the male drones which take part in the mating, far away from the colony, and high in the air.

After the mating ceremony is done, all the drones are killed off by the worker bees so that they did not come back to the hive and grow fat eating honey and doing nothing else.

Around 1760 a beekeeper named Thomas Wildman made the first wooden bar hive. He placed some wooden bars over an ordinary beehive and encouraged the bees to make their combs attached to the wooden bars.

This is just a repetition of the knowledge used by the ancient Greeks, millenniums ago, when they made movable frame honeycombs with the swarms attached to frames which could be moved about by the beekeepers.

The ancient Greeks knew that there had to be a space between combs through which the bees could move about. It was between 5 – 7 mm. If there was this much space between a number of frames, the bees would build honeycombs without attaching the honeycomb walls together with the use of beeswax and propolis.

So that means that the beekeeper could easily slide the honeycomb during inspection without any harm being done to the comb or to the bees. That means the larva, pupae as well as the eggs could be protected in the hive. And the beekeeper could remove the honey gently from the honeycomb and return the empty hive back to its original place.

Many movable honeycomb designs are available all over the earth, depending on the locality, the species of the bees and the availability of flowers during the season. In the USA, you can get Langstroth and Dadant patent designs. In Germany, France, England, Scandinavia, Italy, and Denmark, you are going to get local and national traditional hive designs coming down either from the ages, or by innovative supposedly modern apiarists.

Langstroth model

So what should your hive have? Basically, traditionally, and naturally, a honeycomb would just have a floor and a roof and a protective covering. As one is growing more modern day by day designs include brood boxes crown boards, floors, roofs, and honey supers [area where the honey is stored]. Traditionally hives were made of cypress, pine, and cedar. But nowadays, you are going to get hives made of polystyrene.

The Queen excluder is kept between the honey super and the brood box, so that the Queen does not go and lay her eggs in the honey storage cells. Also, in the 20th century with a number of mites and parasites attacking your honeycombs, removable trays are used instead of hive floors. Wire meshes are also used, to prevent the entry of a parasite into the hive.

Bee stings

Bee stings have very rarely been known to be lethal. However, there have been cases known, where an immediate allergic reaction to a bee sting has caused danger to the life of the victim. It is said that if you find yourself stung once, and survive, you are going to develop an antibody and that is why a number of beekeepers get themselves stung purposely, every season! Funny!

However, it is known that beekeepers have an large number of antibodies which prevent the venom of the bees from taking effect upon the skin and the system of a human being. The protective clothing is going to protect you from the bee sting, but if you have been stung, carefully remove the stinger making sure that the glands full of venom do not squeeze in the process. A sharp fingernail can do the scraping and it has been in use for millenniums traditionally and naturally and instinctively!

To make sure that any sort of venom does not spread in the affected area, you may want to apply a little bit of honey to the stung portion after the sting has been removed to prevent any sort of infection too. And then when you go inside the house, take out some ice from the fridge, and apply the ice on that area to neutralize the venom even further.

Beekeeping in Your Backyard

In ancient times it was always believed that the beehive should not be interfered with, until it is time for you to harvest the honey. However as we are working in modern times, and we think that our knowledge is always so superior, we cannot resist inspecting the hives frequently and often, giving the bees some routine medication, feeding them with sugar water spraying them with water, especially in the summer, and any other stunts which somebody has advised to us, and we think is beneficial for the good health and cheer of the bees. I have seen all this being done, especially from first time beekeepers were so excited with their new hive that they have to go and see what the bees are doing opening the hive, inspecting the honeycomb, watching the cells being made and interfering with any normal bee activity throughout the day. Poor little things. Leave them alone!

Just tell me how would you like it if some large hulking creature keeps hovering over your shoulder when you are cooking the dinner saying things like my my, this is so interesting, how do you know about the amount of spice to be put in the food and is this the proper way to chop up the vegetables, and this is a thing I did not know, how to stir the soup.

After 15 minutes he goes away and then comes back again about half an hour later with a new set of fool questions, getting in your way and apologizing when you want to do your work. The bees are really patient. They suffer fools gladly. Human beings would have lost their tempers and patience within half an hour and conked such a mutt upon his head with a cooking ladle and throw him out of the kitchen.

So leave your bees alone and do not go inspecting them every week, because you want to see how big the hive has grown.

Once I asked my uncle why his bees preferred his own homemade hives to swarm in every year, and he said that he had made the garden so attractive with a large variety of flowers in different colors, bright white and yellow that the bees had decided that this was a really nice place to be, no pun intended. Apart from that, his apple orchard harvest had turned out better because of the quick pollination of the flowers by the bees in his own garden. And as his garden was in bloom 24/7, 356 days of the year, no wonder bees, wasps, bumblebees, and all sorts of insects kept buzzing around the place in a drowsy monotone which was capable of inducing sleep in any human being who went into the garden and got intoxicated on the smell of the flowers.

Apart from eucalyptus honey, he also had mustard honey depending on the flowering season of the mustard. And people used to come from miles around requesting him whether he could supply them with about 20 bottles, at a time because word-of-mouth had spread that he had hundred percent honey. Demand always exceeded supply at the rate of 5:1! Within five years he decided that beekeeping was a much more responsible problem undertaken as a business done by two – he and aunt. With Oscar as approving audience. So he gave away his honeycomb frames and hives to a cottage industry after training the villagers to keep the bees using methods he had been using through experience and through traditional advice.

But before that he had taught me how to recognize the three different types of bees – the Queen bee, the worker bee, and the drone. There would be only one Queen bee which would be the mother of the whole brood and capable of laying eggs. The worker bees were female who also did the guard duty in the hive and raised the baby bees/larvae hatched from the eggs. These were anywhere between 50,000 – 100,000, depending on the size of the swarm/hive. The drones were the male bees, numbering many thousands in the springtime and just a few in the winter.

The life span of a queen bee was anywhere between two years to three years. She would spend her lifetime laying anywhere between half 1 million to 1 million eggs starting from late spring to the summer. A queen bee was normally chosen from a female worker egg but instead of being fed honey, that larva would be fed royal jelly. This would bring about a great metamorphosis in the body structure and the growth rate of that particular worker bee who is the Queen. Incidentally it is the queen who is responsible for the production of a number of pheromones which prevents the worker bees from laying any eggs in the hives. That is because their ovarian development is suppressed due to the production of these pheromones.

The Mating Flight

This queen bee which has been fed on royal jelly for the last 15 days since it was hatched is going to emerge from her Royal cell. She is going to stay in the hive from anywhere between four – seven days before she takes flight into the air. This is known as the nuptial flight. The first flight can be considered an orientation marking out the position and placement of the hive in the garden. After that, more flights are going to be made lasting anywhere between 10 minutes to half an hour. The Queen bee is going to be followed by a number of drones and innumerable matings with a number of drones is done during these flights. During the nuptial flights the Queen bee manages to store a very large amount of sperm which is going to be capable of fertilizing millions of eggs.

If this nuptial flight is not successful due to a number of factors like possible bad weather or even if the Queen bee is incapable of coming out of her cell, the hive's existence is jeopardized. The workers know that such a queen bee is infertile and incapable of laying fertile eggs. That is why they might kill the Queen bee and begin feeding another worker bee royal jelly. This may sound very cruel but for millenniums it has always been natures way of survival of the fittest.

When I asked uncle why these flights took place so far away from the hive and so far above, in the air, he said very insouciantly in his Commando army parlance "when the going gets tough, the tough get going." He explained that this was natures way to separate the top-quality from the weak duds and only the strongest as well as the fastest drones could mate with the Queen and pass on their strong gene line. Very logical and sensible answer I thought. And true.

The life span of a worker bee is about one month to 5 to 6 weeks. That is because it has exhausted itself throughout the summer, working hard for its hive. However, with the coming of the autumn and the beginning of winter when the worker bee can relax a little because it is not raising the brood or harvesting the nectar, its lifespan can extend up to four months or so. If it is lucky, it can even survive the winter.

A worker bee is one of the busiest of all the insects found in the insect kingdom. In the initial days of their life, they are going to do the normal workaday chores in the hive - like cleaning the cells, cleaning up all the debris accumulated in the hive, do lots of housekeeping, feeding the larvae, and then as the days go by, they are going to advance to more responsible duties like feeding the younger larvae. Within two weeks, they are going to learn how to receive the pollen from other bees as well as nectar. They are also going to learn how to make wax in order to repair and build the comb.

As the bee grows older, it is going to be given more responsibility like guarding the hive, and also ventilating it by fanning its wings. An older more responsible bee goes out to gather honey, pollen, water, nectar, and propolis. It may also attack another hive in order to steal the store of honey there.

A Drone's Life

Down the ages, the term "drone" has been used to describe someone or something, which is totally supposedly use less, eating and drinking and enjoying life while somebody else does its work for it. The drone in the hive is about as large as the Queen bee and twice the size of a simple worker bee. It does absolutely no foraging. Its only job is to take part in the nuptial flight. Drones are raised a couple of weeks before the bees begin to build the cells for a queen bee. This is so that when a queen begins to fail, they are going to have plenty of drones as well as a future Queen to keep the hive going strong.

After the nuptial flight is over, the drones are thrown out of their lazy existence in their hive and their wings and legs are torn away so that they die, now that there is no use for them.

It takes 15 days for a queen to develop, 24 days for a drone to develop, and 21 days for a worker bee to develop in the hive.

If you cut away a cross-section of the hive, you are going to see a number of cells in which you are going to find the larvae, eggs, pollen, and honey. The pollen is what is going to be fed to the larvae because it has lots of protein in it. The honey has more energy. The royal jelly is produced by the nurse bees, who have decided which particular egg is going to be the future Queen bee. An ordinary larva is going to be fed pollen and honey. Royal jelly is only for the Queen.

Above the brooding area, the worker bees make a number of cells where they store the honey. These boxes are now called honey supers. That means it is easy for the beekeeper to remove the honey without disturbing the larvae or the brooding nest or the main colony of the bees.

Remember that the bees need to survive in the winter by feeding on the honey. If you steal away all the honey stored, now that the winter is coming in, you are going to starve your poor little bees. That is why you will have to feed them sugar solution or even corn syrup in the autumn so that they have enough food in their stores, so that they can get through the winter.

Impulse to Swarm

Remember that the bees are naturally wild and that is why they want to swarm which means that the bees take flight and there you are, half of the worker bees have gone with the wind. The moment you see them collecting pollen for the feeding of the larvae, depending on the area, this means that the brood nest has begun to expand. Starting in January, it is going to go on up to mid-May. While the flowers are in full bloom, you are going to have lots of harvesting bees.

If you are lucky, there are going to be two flows of nectar, depending on the region and the amount of flowers available, there. One is going to be in the spring. The other is going to be in August. So with a little bit of experience you are going to know when the next nectar flow is going to take place in your land and this is when you need to make sure that your colonies have the maximum number of bee harvesters ready to take advantage of the flowers and the pollen in order to make lots of honey and money for you.

It is going to take a bit of time and experience in order to make sure that the impulse to swarm is checked and a new Queen is bred in the same hive without the bees deciding to make a break for it. So you could almost say that beekeeping is a little bit of trial and error, and lots of experience.

Fertilized eggs laid by the Queens become worker bees. Unfertilized eggs become drones. So if there are lots of unfertilized eggs, you are going to have lots of useless drones in the hive, but the Queen bee uses her natural instinct. If she sees a small cell, she is going to lay a worker bee egg. A larger cell is going to have a drone egg placed in it.

It is the job of the worker bees to make the cells, depending on their instinctive need for what they want, more workers or some more drones because a new Queen bee is being readied for the hive. As the queen grows older, she is going to run out of sperm and begin laying more drone eggs. That is when the bees decide that a younger Queen has to be prepared. And so they begin feeding royal jelly to a worker bee.

That means that they begin to make queen Bee cells from worker cells. Bees normally follow two different types of social behavior. One is superseding the existing Queen and replacing her without any sort of swarming. Or they might decide to start up another swarm by making swarm cells. That means the hive is going to be divided into two different colonies.

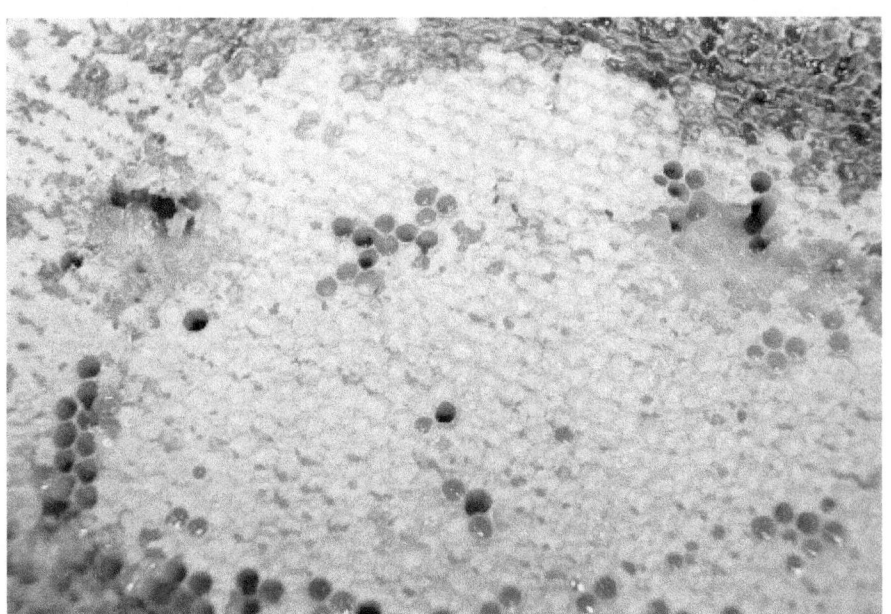

Royal Jelly

One does not know whether the bees may decide to swarm and start up a new colony or supersede the Queen in their old existing colony. For example, Apis mellifera shows different types of behavior in different types of the World, depending on the weather availability of the food and atmospheric eco-types available in the area.

Supersedure as a behavior trait is more preferred in bees because that means that they are not going to do any swarming and no bee stock is going to be lost. They are just going to get a younger Queen to take the place of the older one. One or two new queen cells are made right in the center of the brooding comb. And the worker eggs laid there become the future Queens.

However, when swarming is done, they are going to be a number of queens, because a number of different hives are going to be created following all these young Queens in the brood comb. These cells are normally made all around the edge portions of the comb, and the top and the bottom cells are modified into queen cells.

The moment the new Queen cell is hatched, the old queen is going to leave her hive. A number of young worker bees are going to accompany her to make a new hive colony. But before that, scouts are sent out in order to look for suitable places where the new hive is going to be made. These are going to include crevices and hollows. The moment a suitable place is found, the whole new swarm is going to set up house there. And within a couple of hours, you are going to have new cells and combs which are going to be filled up with honey. This honey was eaten by the worker bees, before they left to accompany their old queen on her new adventure.

Only a young bee can produce wax by stroking some cells on its stomach, and that is why a swarm is going to have a large number of young worker bees. A number of virgin young queen bees are also going to accompany the Queen Mother. And after their nuptial flight and after they start laying eggs, the Queen bee is replaced and retired. Or she may be superseded in her new colony.

So how are you going to know whether your bees are ready for swarming or not? Bees normally do not swarm until all the brood combs cells are filled up with either eggs or larvae. This is going to occur in the late spring, especially when the other parts of the honeycomb is filled up with honey.

The moment the Queen thinks that there is no space for her to lay any eggs, and the hive is becoming overpopulated, she is going to give the order for the division of the population. Half of the bee population is going to go with her to set up a new colony. The old colony already has a number of baby larvae which are going to populate the previous hive under the guidance of the nursing bees and guarded by the worker bees.

A new hive is definitely not going to have any eggs or larvae. However, thanks to a number of energetic worker bees, making large number of combs for a new brood, you are going to see a new hive coming up before your eyes. Within hours, a new honeycomb is going to be set up.

Remember that if the honeycomb has a queen, less than a year old, you may not find any swarming in that particular comb. However, if the honeybee Queen is more than a year old, her instinct to swarm is greater, and that is why the moment the honeycomb gets a bit too crowded, you can know by experience that your bees are ready to swarm.

This is normally going to be done in the early spring and that is why a number of experienced beekeepers keep a careful eye upon the cells. The moment they see some queen cells appearing, they know that the bees are ready for a division of the hive and swarming.

They are going to be anywhere between 12 to 15 queen cells. After being fed for eight days on royal jelly, the Queen is going to emerge from her pupa within the next week. During this time the worker bees eat a lot of honey, so that they can produce more honeycombs in their new hive.

These well fed worker bees are not so aggressive and that is the reason why beekeepers find them easier to handle. For about 24 hours, you can handle them without the need of a veil or gloves.

A major prime swarm is going to have about 10,000 bees. The moment a new Queen emerges from herself, she is going to look out for other yet unborn queens, and seek to kill them. However, she is prevented in doing that, by the Guardian bees and that is why she takes some more workers to set up another swarm.

An experienced beekeeper is normally on the lookout for a prime swarm, but if he can get any swarming with a queen bee looking for shelter, he can capture that swarm and entice it into new hive boxes. Otherwise, the bees are going to take flight and become wild taking shelter in any sort of natural hole like hollows, empty spaces, and any place where they can set up their home.

If you are unable to stop your bees from swarming and the population has been reduced drastically, you need to add a number of frames with lots of larvae and eggs to the hive. This is going to populate the hive and replenish it. It is also going to give the hive another opportunity to raise another Queen bee.

Artificial Raising of Queens

A colony cannot do without a queen bee

This is normally done when there is no queen in the colony. As the survival of the colony depends upon the Queen, because she is the only bee capable of laying eggs, the worker bees immediately begin making queen cells of some cells where eggs less than three days old have been laid. You can call these Emergency Cells, because these bees are going to be fed royal jelly, pollen, and honey.

The ovaries of these queen bees are developed fully and they are the only ones capable of laying eggs.

This capacity of raising a new Queen by the worker bees is taken full advantage of by the beekeepers who go through a procedure called splitting up the colony.

They are going to take a number of brood combs. The old queen has left behind. Those combs are going to have larvae and eggs which are less than three days old. They should also have a number of nursing bees. These bees are going to take care of the brood and keep the hive warm.

These new combs are now going to be placed in a place where there are other pollen and honey containing "nucleus hives."

The moment the nursing bees notice that they do not have a queen, they are going to make emergency cells to raise a new Queen.

Just like other insects in the insect kingdom, the bees are also prone to a number of diseases which include diseases caused by viruses, bacteria, fungi, parasites, protozoa, and poisons. Thanks to the global ease of transportation of diseases, many bee colonies and thus industries have been wiped out in the early 2000's, and that is why nowadays beekeepers are very careful about where they get their bee stock from.

The top producers of honey in the world are Russia, Spain, Hungary, Romania, and Germany, followed by Serbia, Bulgaria, Greece, Romania, and France, in Europe. In Asia, China and Turkey lead, followed by Iran, India, and South Korea. In Latin America, Argentina, Mexico, and Brazil are the top producers of honey while in Africa, Ethiopia, Tanzania, Angola, and Kenya, followed by Egypt have been producing honey traditionally for millenniums.

Conclusion

Now that you know a little bit of honeybees and how you can set up your own beekeeping yard in your own backyard, here are some very important tips which you cannot lose sight of.

Bees need plenty of shade. They also need plenty of water and lots and lots of pollen and nectar sources. That means there should be plenty of garden area in abundance so that the bees do not have to fly more than 5 miles away in order to feed themselves copiously.

That is why if you have a really nice colorful flowering garden, lots of trees full of fruit, – let us say an orchard, – a water source with plenty of water, you are already to set up beekeeping. There are a number of popular bee plants which are going to encourage the presence of the bees in your hives.

This URL is going to give you more information on the bee- plants which you can grow in your garden.

http://www.gardenersworld.com/plants/features/wildlife/plants-for-bees/1107.html

Beekeeping is best done by the side of a river, as they love that abundant water source. Or second best, next to streams with plenty of flowing water, especially in trees with lots of shade.

Vegetation normally blooms in September and October in lower altitudes. This blooming takes place between February and April in higher altitudes. You as beekeepers should know the bloom time periods of your location so that you can situate your hives in both areas for a higher yield of honey.

Honeybees, as I told you need lots of water, and that is why in the dry season, my uncle did not want his bees to go flying to the little rivulet about 100 yards from his garden because he did not consider the water to be really clean. He just put out lots of water containers and also made sure that his fountain never ran dry.

And to prevent the bees from drowning themselves, the ledges of the fountain in which the water droplets dropped had lots of twigs, straws, stones, and other bits of debris on which a bee would cling, when drinking its fill of cool water in the summer.

Harvesting Time

So how do you know about the right harvesting time? That is when the bees begin to stay outside the hives, instead of going in. They may also stay near the exit because all the cells have been filled up with lots of honey.

You are going to learn through experience the right time for harvesting the honey. This is normally done when the flowering season has ended. That means the bees have made enough honey and are now protecting it, they have covered it with a fine white layer of beeswax.

The keeper now knows that the honey is ripe. He is first going to inspect the combs, which are nearer to the entrance and outside of the comb. This is where the honey is going to be placed in the first instance. In ancient times, the beekeepers just cut the honeycomb and sold them in the local markets and it was your job to remove the honey, and the beeswax. Today, these honeycombs are being sold at premium prices.

Honey extractors are specialized equipment, which are going to help you extract honey, but traditionally, filtration was done through fine Muslin cloth after the honeycomb was broken, placed in the muslin and a heavyweight placed upon the muslin and the honeycomb so that the honey could drip, underneath into a container. After the honey was obtained, the beeswax was gathered by just placing the remainder of the honeycomb in a double boiler where the honeycomb containing utensil was placed in another utensil full of boiling water. Slow heating would melt the beeswax. After that you could either purify the honey into a block or use it in its un-purified shape, where it would have some honey in it and other impurities.

Nowadays, people extracting large quantities of honey normally spin the frames in extractors using a centrifugal motion and principle. After that, the empty honeycombs are going to be placed back in the hive again. This honey spinner was the invention of an Italian army officer, in 1865. He just supported the combs in a framework made of metal, and then allowed the combs to be spun around inside a container. The honey would be thrown out, thanks to centrifugal force, and then the honeycomb could be placed back in the hive without any harm done to the honey bees. This revolutionized the honey production industry through really efficient harvesting of honey. People wonder why the yield of beeswax in frame hives is lesser than that of traditional hives. The reason is that the top priority of the Bee is to produce honey and not beeswax because it already has a "recycled" comb.

Honey extraction by centrifugation, then impure honey is screened

Traditionally in Kenya, smokers are made up of Acacia twigs – Acacia mellifera or Acacia seyal. These are called *Kinyua* or *Muthiya* in the local dialect. For millenniums, these twigs have been used because they produce lots of smoke and also keep lit for longer periods of time.

Incidentally, why these twigs have been used down the centuries in Kenya is because honey harvesting is normally done at night by totally un-garbed harvesters. They say that any protective cloth is going to entangle the bees. Thus, the chances of their being stung is going to be greater because the bee is going to be thoroughly confused by the material, preventing it from flying away and is going to use its natural defensive mechanism – the sting.

Indigenous Peoples Tips for Attracting Bees

When I began to do more research on beekeeping, I found that traditionally there have been plenty of ways and methods in which the beekeepers enticed a swarm into a hive down the ages. I have already given you the example of rubbing the old hive with molten beeswax mixed up with scented twigs of Oscimum – the basil family.

You can either put the twigs in the hive or you can just light them and allow the smoke to permeate all through the area, purifying it as well as enticing all the insect life in the vicinity! The reason is that oscimum has a pheromone which is equivalent to a brood pheromone for the honeybees

In many parts of Asia as well as in Uganda, where there is plenty of lemongrass, the inside of the hives are rubbed with leaves of fresh lemongrass. In Kenya, animal fat, mixed with honey, beeswax and pollen was rubbed inside the hive, traditionally so that bees were attracted by that amazing scent and followed it.

Incidentally, the same Kenyan friend of who I talked a little bit on the introduction, told me about an indigenous practice which was practiced many centuries ago by her people who collected the honey at night. They covered their whole body with a paste of river mud, thus masking their own human scent while protecting their skin. By the time the river mud layers were ready to dry and had begun to peel off, the harvesting was done!

These harvesters also had long knives with them in order to break up the honeycombs while killing all the snakes which lived in the holes next to the bees. How on earth would the honey guide lead you to a snake hole, as a reward for not feeding it if it did not know which hole held a mamba or any other horrible poisonous snake?

If you find the cells with a layer of wax on them, the honey is ripe. It is called capped honey. You are not going to mix up un-capped honey with capped honey. An experienced beekeeper is not going to touch a honeycomb with the brood combs and cells in it. He is also going to wait till the unripe honey is capped before he harvests it.

For millenniums, honey harvesters have known that if you do not put the lid on the harvested honey, it is going to ferment because it has absorbed the moisture present in the air.

In Kenya, Egypt, and Ethiopia, this fermented honey drink was part of the ancient cultural tradition to be used on festive occasions, until this tradition was stopped by these European missionaries – especially from Britain, where beekeeping was relatively unknown, when compared to other European countries, in the 18th – 19th century who considered this tradition to be barbaric and even "Unchristian like."[2] This particular information bore out what was told to me by Kanika.

And that is why these ardent, enthusiastic people stopped honey bee culture and growth in many parts of Africa just because any old tradition and culture, about which they did not know, was a heathen practice, according to them. And that is why the Christian converts were told to give up beekeeping, which they did, especially in the 19th century and in Victorian times when every ship from Europe going to East Africa had about 8 to 10 missionaries. And each with his/her own interpretation on how these ancient people could be civilized. And so the knowledge of centuries was lost, but that has always been the history of mankind with people giving in to more forceful persuasions, all for the sake of keeping the peace.

However, in the 21st century, a number of organizations starting up in East Africa has again begun collecting all this indigenous knowledge and encouraging the beekeeping industry again in those areas.

So now that you have got lots of information on bees, is not it time you started up your own little bee industry in your own backyard?

If you are interested in knowing more about the miracle of honey, you may like to read my best-selling book – The Miracle of Honey, found here:

http://tinyurl.com/z2oza2p

Live Long and Prosper!

---

[2] I got this particular information from a Kenyan beekeeping related website. This surprised me, because I had never heard of this sort of behavior from Asian missionaries. They never prevented bees from being kept by the natives in British colonies in Asia, because beekeeping is still very much a part and parcel of life in the British colonies in Asia today.

Author Bio

Dueep Jyot Singh is a Management and IT Professional who managed to gather Postgraduate qualifications in Management and English and Degrees in Science, French and Education while pursuing different enjoyable career options like being an hospital administrator, IT,SEO and HRD Database Manager/ trainer, movie , radio and TV scriptwriter, theatre artiste and public speaker, lecturer in French, Marketing and Advertising, ex-Editor of Hearts On Fire (now known as Solstice) Books Missouri USA, advice columnist and cartoonist, publisher and Aviation School trainer, ex-moderator on Medico.in, banker, student councilor ,travelogue writer … among other things!

One fine morning, she decided that she had enough of killing herself by Degrees and went back to her first love—writing. It's more enjoyable! She already has 48 published academic and 14 fiction- in- different- genre books under her belt.

When she is not designing websites or making Graphic design illustrations for clients , she is browsing through old bookshops hunting for treasures, of which she has an enviable collection – including R.L. Stevenson, O.Henry, Dornford Yates, Maurice Walsh, De Maupassant, Victor Hugo, Sapper, C.N. Williamson, "Bartimeus" and the crown of her collection- Dickens "The Old Curiosity Shop," and "Martin Chuzzlewit" and so on… Just call her "Renaissance Woman" - collecting herbal remedies, acting like Universal Helping Hand/Agony Aunt, or escaping to her dear mountains for a bit of exploring, collecting herbs and plants, and trekking.

Check out some of the other JD-Biz Publishing books

Gardening Series on Amazon

Download Free Books!

http://MendonCottageBooks.com

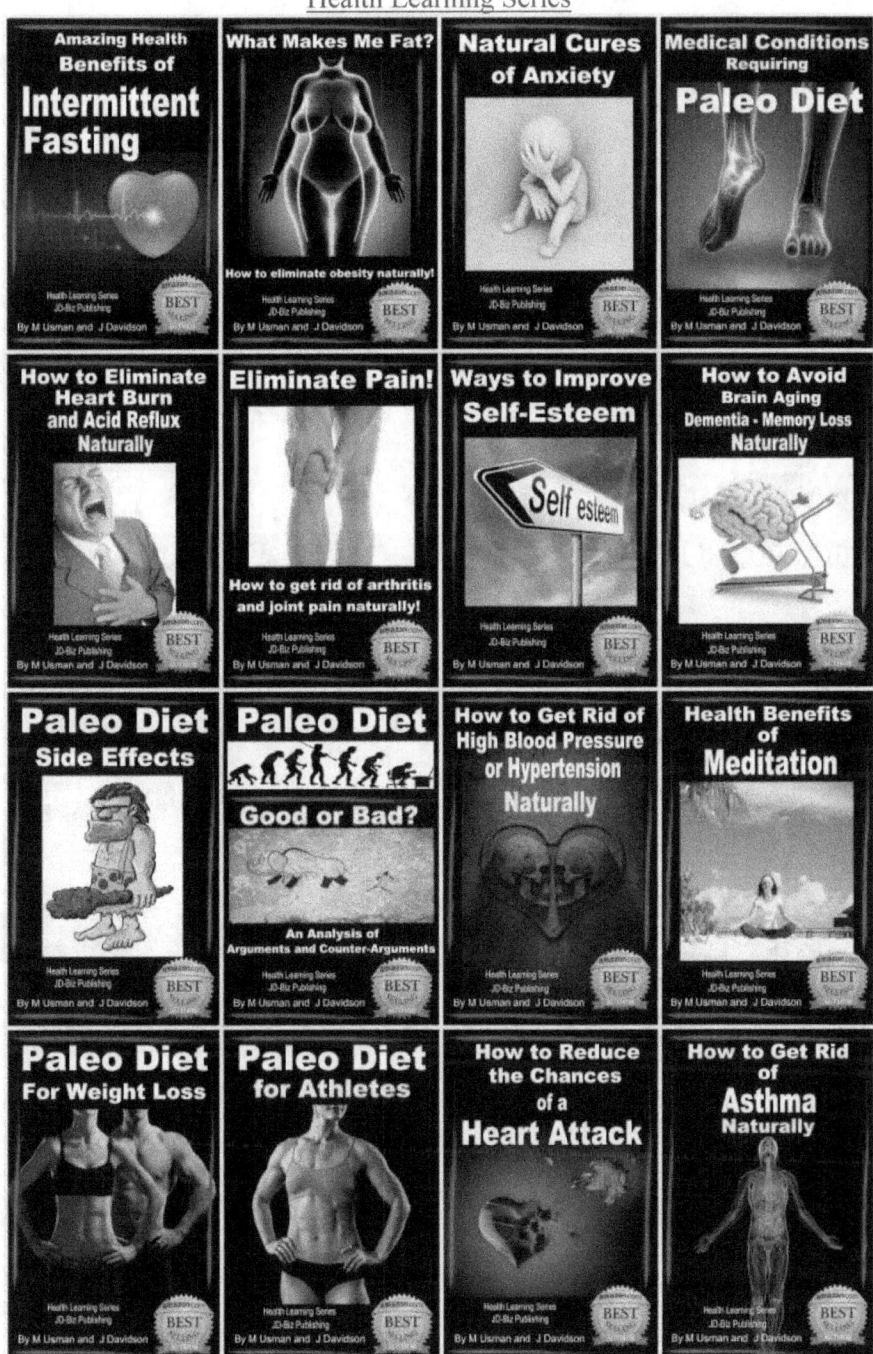

Amazing Animal Book Series

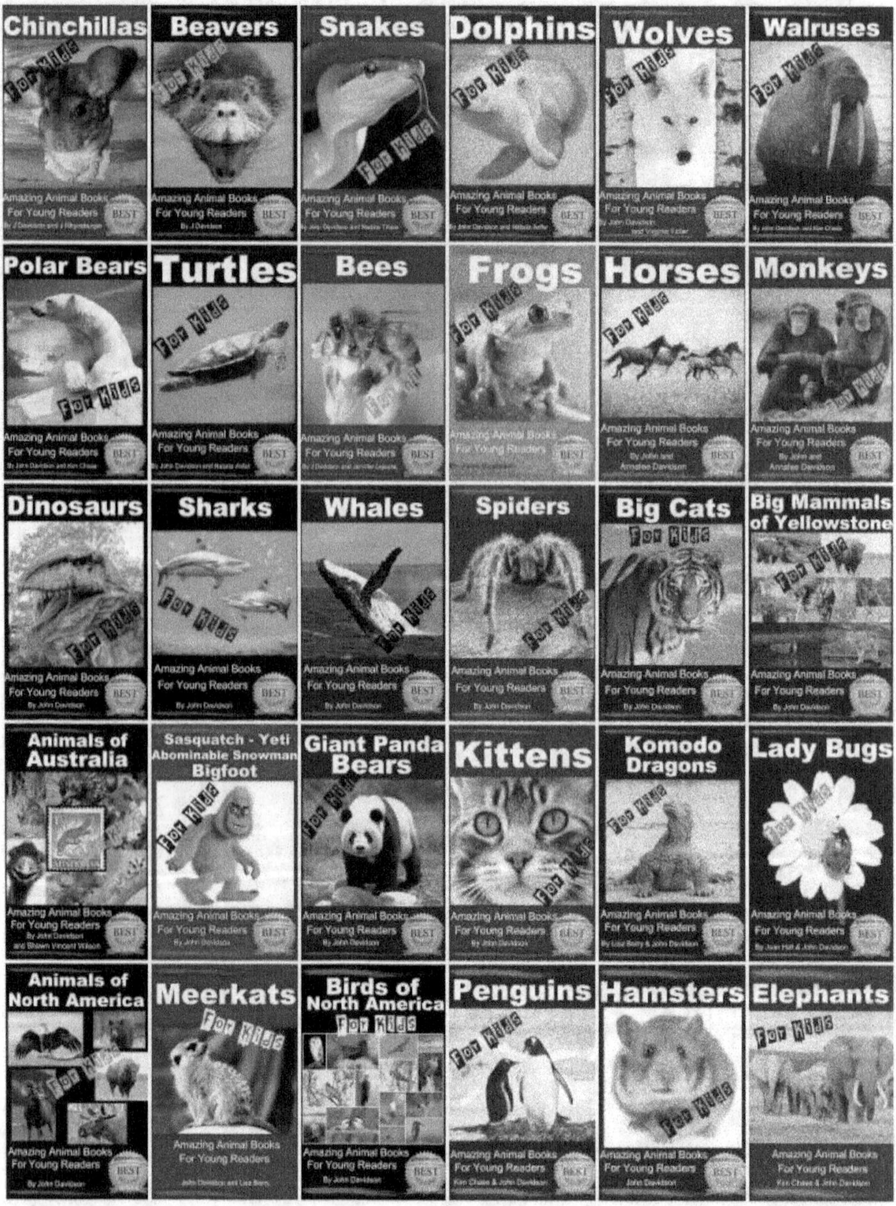

Learn To Draw Series

How to Build and Plan Books

Entrepreneur Book Series

Our books are available at

1. Amazon.com

2. Barnes and Noble

3. Itunes

4. Kobo

5. Smashwords

6. Google Play Books

Download Free Books!

http://MendonCottageBooks.com

Publisher

JD-Biz Corp

P O Box 374

Mendon, Utah 84325

http://www.jd-biz.com/

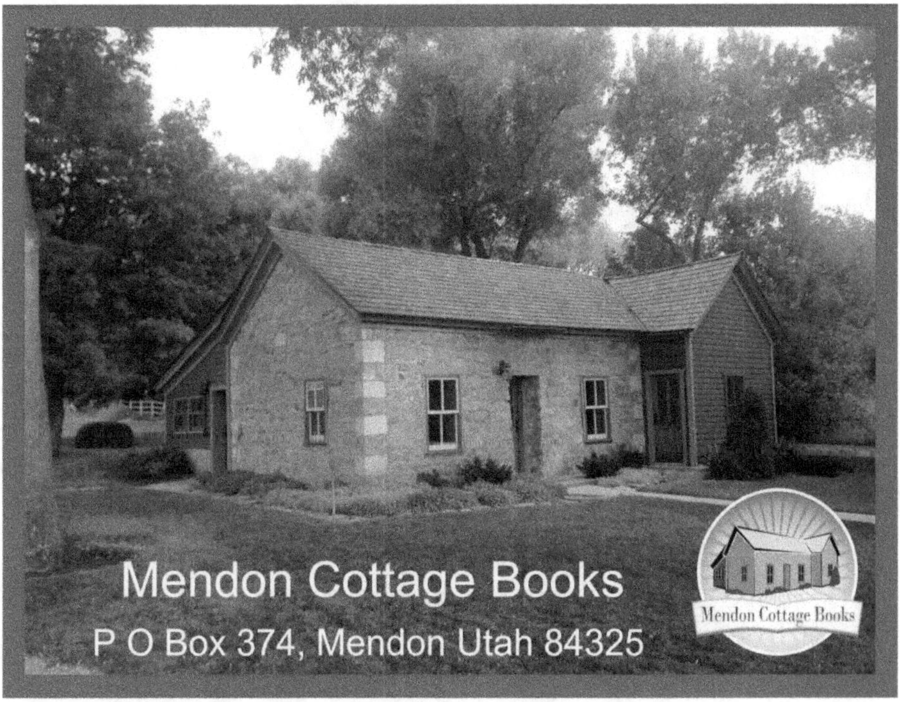

Mendon Cottage Books

P O Box 374, Mendon Utah 84325

Mendon Cottage Books